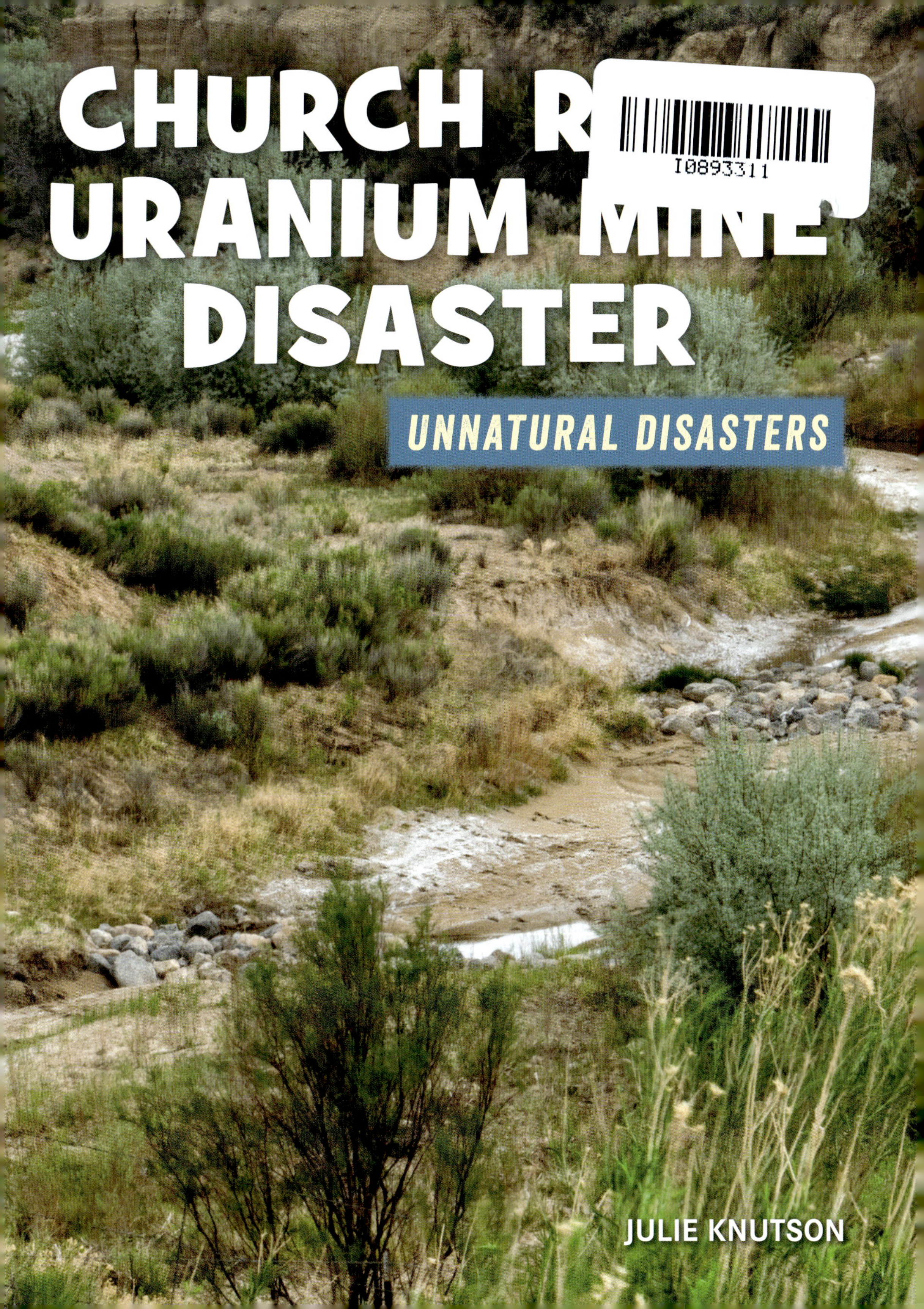

CHURCH R[illegible] URANIUM MINE DISASTER

UNNATURAL DISASTERS

JULIE KNUTSON

CHERRY LAKE PRESS

Published in the United States of America by Cherry Lake Publishing Group
Ann Arbor, Michigan
www.cherrylakepublishing.com

Reading Adviser: Marla Conn, MS, Ed., Literacy specialist, Read-Ability, Inc.
Photo Credits: © Barbara Ash/Shutterstock.com, front cover, 1; © REUTERS/Alamy Stock Photo, 5, 26; © Courtesy the Library of Congress, LC-USZ62-101174, 9; © UncreditedAP/Shutterstock.com, 10; © 12 Smith Archive/Alamy Stock Photo, 12; © Everett Collection/Shutterstock.com, 14; © Jim West/Alamy Stock Photo, 17; © AP/Shutterstock.com, 19; © Smooth Ninja/Shutterstock.com, 20; © Mary Hudetz/AP/Shutterstock.com, 23; © sumikophoto/Shutterstock.com, 24

Cherry Lake Press is an imprint of Cherry Lake Publishing Group.

Library of Congress Cataloging-in-Publication Data
Names: Knutson, Julie, author.
Title: Church Rock uranium mine disaster / Julie Knutson.
Description: Ann Arbor, Michigan : Cherry Lake Publishing, 2021. | Series: Unnatural disasters : human error, design flaws, and bad decisions | Includes index. | Audience: Grades 4-6 | Summary: "Human modification of the environment always carries a risk of accident and folly. Explore the causes and consequences of the Church Rock, New Mexico uranium mine disaster of 1979. Guided by compelling questions such as, "What led to this disaster?," "Who was impacted by it?," and "What changed in its aftermath?" the interdisciplinary content blends social studies and science. Ultimately, it pushes students to consider how humans can meet their need for resources in a safe, sustainable way. Books include table of contents, index, glossary, author biography, and timeline"–Provided by publisher.
Identifiers: LCCN 2020040021 (print) | LCCN 2020040022 (ebook) | ISBN 9781534180161 (hardcover) | ISBN 9781534181878 (paperback) | ISBN 9781534181175 (pdf) | ISBN 9781534182882 (ebook)
Subjects: LCSH: Uranium mines and mining–Accidents–New Mexico–Church Rock–Juvenile literature. | Tailings dams–Accidents–New Mexico–Church Rock–Juvenile literature. | Dam failures–New Mexico–Church Rock–Juvenile literature. | Radioactive wastes–Environmental aspects–New Mexico–Church Rock–Juvenile literature. | Disasters–Social aspects–New Mexico–Church Rock–Juvenile literature. | Navajo Indians–New Mexico–Juvenile literature.
Classification: LCC TN319.5 .K58 2021 (print) | LCC TN319.5 (ebook) | DDC 363.17/990978983–dc23
LC record available at https://lccn.loc.gov/2020040021
LC ebook record available at https://lccn.loc.gov/2020040022

Cherry Lake Publishing Group would like to acknowledge the work of the Partnership for 21st Century Learning, a Network of Battelle for Kids. Please visit http://www.battelleforkids.org/networks/p21 for more information.

Printed in the United States of America
Corporate Graphics

ABOUT THE AUTHOR

Julie Knutson is an Illinois-based author. In her spare moments, she enjoys investigating new places and ideas alongside her husband, son, and border collie.

TABLE OF CONTENTS

Scan newspaper headlines from July 17, 1979, and you'll probably learn that a massive tornado touched down in Cheyenne, Wyoming, the morning before. You might also see that Nicaragua's iron-fisted **dictator** resigned and fled to the United States. Or you might read that President Jimmy Carter pledged $142 billion to energy independence. But there's one news report that you're unlikely to come across. The largest spill of **radioactive** waste in U.S. history—the largest in the history of the world at that time—had happened at 5:30 a.m. one day earlier.

The Northeast Church Rock **uranium** spill occurred on Navajo land in rural New Mexico. Uranium was mined at this site—and more than 500 others—across the Navajo Nation. It was used to provide fuel for nuclear **reactors** and to build military weapons. A dam held back 94 million gallons of radioactive water until it broke, pouring the water and 1,100 tons of solid waste into the nearby Rio Puerco. Water samples showed that radioactivity measured *7,000 times* the safe limit.

Uranium mining in Church Rock, New Mexico, began in 1968.

On the morning of July 16, residents awoke to hear rushing water. To some, it "sounded like a thunderclap." Others later told documentary filmmakers that the river—key to supporting life and livestock in the region—suddenly "smelled like rotten food." They noted that "white particles [were] clinging to the river banks, as if it had just snowed." This toxic water snaked through Navajo lands in New Mexico and into Arizona. It seeped into the ground and impacted people far downstream.

Human settlements tend to rise up along the banks of rivers—the Tigris and Euphrates in Mesopotamia, the Nile in North Africa, and the Yangtze in China. That's no accident. Water is essential to human life. For the Navajo, the Puerco provided water in a barren desert environment. These waters were critical to the survival of people, plants, and livestock.

At the time, this environmental catastrophe scarcely received mention outside of the southwestern United States. Even today, it is less well-known than another accident that happened the same year, the partial **nuclear meltdown** at Three Mile Island in Pennsylvania, which released less radioactive waste. Why?

The events at Church Rock were decades in the making. The mine and mill operator, United Nuclear Corp (UNC), knew about cracks in the dam that ruptured. Today, members of the Navajo Nation still live with the effects of the spill. Why were people mining for uranium in the first place? How did this accident happen? Who was held responsible? What was done to restore the health and safety of this land and the people who call it home?

A String of Unnatural Disasters

In the late 1970s, the United States was rocked by a series of environmental catastrophes. In 1978, the world's attention fell on Niagara Falls, New York. The city's Love Canal neighborhood had been built on top of a toxic waste dump. The sludgy residue of these buried chemicals bubbled up in backyards, schoolyards, and basements. Residents—who weren't told this when they bought their homes—suffered **chronic** health conditions and fretted over the well-being of their kids. Activists worked tirelessly to shine a spotlight on environmental injustice. They demanded that officials hear their plight and fix the situation. In the process, they launched a new chapter in the environmental movement.

The next year, on March 28, 1979, a partial meltdown at Three Mile Island nuclear power plant near Pennsylvania's capital threw the country into crisis. The accident catapulted questions about the safety of atomic energy to the top of public debate.

Before

In Navajo legend, those who disturb the earth ask for trouble. Any mineral riches buried below ground are meant to stay there. If humans dig and mine and drill, they will bring powerful monsters to the surface. Those restless monsters will haunt them. They will torment the land and future generations of people.

While Navajo *is more common,* Diné *is the name that many members of the Navajo Nation prefer. Pronounced "Di Nay," it means "the people." The term comes from the Nation's own language, while* Navajo *comes from a Tewa Pueblo term meaning "place of large planted fields."*

The Navajo people call themselves *Diné*. In their own language, this translates to "The People."

The Navajo descended from nomadic people who migrated from the areas that are now Canada and Alaska. They have called the Four Corners area of the Southwest—where the boxy states of Arizona, New Mexico, Colorado, and Utah meet—home since the 1300s. As the country's largest Native American community, the Navajo Nation numbers 180,000 people. Its area outsizes 10 U.S. states.

From 16th-century Spanish conquistadors to Anglo settlers in the 1800s, outsiders have long eyed Navajo lands. Precious metals and ores were sought in the desert rocks. In 1864, Navajo

From 1968 to 1982, the Church Rock Mine extracted around 3.5 million tons of uranium ore.

communities were forcibly removed from their lands by the U.S. government on what is known as the Long Walk. While many Navajo put up fierce resistance, U.S. troops burned villages and slaughtered livestock to force their surrender. The Navajo were then marched to camps 250 to 450 miles (402 to 724 kilometers) away. The government's plan was to "civilize" and **assimilate** the Navajo. During the 4 years that the Navajo were held in government camps, **prospectors** scoured the land for silver and gold.

Decades later, white prospectors looked to the land for oil and coal. Others came in the 1910s to search for **radium**-containing ores for use in everything from cancer treatment to cosmetics. Then, during World War II, something else was found—a key **element** that could be used to build secret military weapons.

At the time, U.S. scientists were scrambling to beat Germany in constructing an atomic bomb. The essential ingredient was uranium, a highly unstable element that releases radioactive energy as it decays. Navajo land held large deposits of uranium. Navajo workers were hired to work in freshly tunneled mines. Because of the top-secret nature of the project, laborers were simply told they were looking for useful metals to help with the war effort. In their contracts, mining companies promised they would take steps to preserve the property and the health and safety of workmen.

The sovereignty of the Navajo Nation—or the power for them to self-govern—was recognized in 1868, after the Navajo returned from the "Long Walk" to their ancestral land. This allowed for self-governance. Each year, April 27 marks Navajo Sovereignty Day.

Miners used Geiger counters to detect uranium ore veins.

But the miners weren't told about the potential risks of the work. Dust from mine explosions—often containing radioactive particles—floated through the air. This dust settled on plants and animals and in the water. Still, mines continued to crop up, forever changing the landscape and disturbing the land.

In the late 1930s, scientists discovered a process called **nuclear fission**. Fission involves bombarding a uranium atom to split it. The process that followed caused a **chain reaction** of explosive energy. Unlocking the uranium was a complex process. First, uranium-containing rocks identified with **Geiger counters** were mined. Then, they were milled. This involved crushing them to ores and treating them with chemicals to extract pure "yellowcake" uranium.

After years of work at secret labs and mines across the country, a bomb was ready to be tested on July 16, 1945. The test site? The New Mexico desert. Upon explosion, a massive mushroom cloud

Navajo Code Talkers

In wartime, communications often get intercepted. To keep plans secret, military agents **encrypt** messages in code. This allows them to conceal movements and tactics.

Ironically, the Navajo language—which the U.S. government tried to stamp out with boarding schools and forced assimilation—served as a key intelligence tool during World War II. The language was neither written nor widely spoken, so enemy soldiers would not know it. Bilingual soldiers were recruited to transmit messages in Navajo. More than 400 code talkers served on the Pacific front, transmitting more than 800 messages during the Battle of Iwo Jima alone. The code was never cracked.

Because their work was **classified**, it wasn't until 1968 that the efforts of these soldiers were recognized. In all, 44,000 Native Americans from tribes across the country—out of a total population of 350,000—served in World War II.

This photograph was taken 9 seconds after the detonation of the atomic bomb at the test site in New Mexico in 1945.

ballooned over the landscape. While the test location was chosen for its "remoteness," **fallout** spread downwind across native lands, impacting rural residents for years to come. Less than one month later, the full power of the bomb was released—first over Hiroshima and then over Nagasaki, Japan. These bombs killed 100,000 people in each city in a single strike. The effects of radiation continued to be felt in the decades after the war.

The atomic bomb brought about the end of World War II. But the pursuit of atomic energy didn't end there. After the war, the United States and its new enemy, Soviet Russia, each sought to stockpile nuclear weapons. A new period was born: the **Cold War**. Government and corporations were also in hot pursuit of peacetime uses for nuclear power. A uranium mining boom continued in the Southwest.

Warnings of the health impacts of uranium exposure were published as early as 1952 and were documented well before that in Europe. But few protections were provided to Navajo workers. Safety measures were lax. Waste wasn't controlled. This impacted everyone and everything in the surrounding area. By 1960, doctors tracking the health of miners had already found an alarming increase in lung cancers and heart disease. While experts made suggestions to improve working conditions—better ventilation, respirators, and dust containment—they went ignored.

Although private companies ran the mines, the U.S. government was the only customer for Navajo uranium until 1971. While it was first used to make weapons, uranium was later used as fuel for the growing nuclear power industry.

CHAPTER 2

The Accident

In 1968, UNC began operating the mine at Northeast Church Rock. Elevators plunged miners 1,600 feet (488 meters) below ground. They worked 24 hours a day, equipped with **dosimeters** to track radiation exposure. Nine years later, the corporation began operating an on-site mill. Now, all of the processing—from raw mineral to yellowcake—would happen in a single location. At the plant's dam, **tailings** were heaped into piles. They were then covered with water to prevent the emission of radon gases and toxic dust from drifting off site.

The Church Rock uranium mill closed in 1982. To this day, the site remains abandoned.

There were warning signs of a possible accident. The earthen dam was quickly overfilled. It was also built on a flood plain. The soil was unstable and shifting. Shortly after the dam was built, cracks were noticed in its walls. Soon, they grew to the size of a human fist. UNC patched them with a clay called bentonite, but this only made things worse. The bentonite chemically reacted to the highly acidic waters in the dam. The cracks grew deeper, yet waste continued to be dumped.

Meanwhile, the state didn't regulate the mine. From the very beginning, officials failed to monitor construction. They didn't know that only half of the required safety material was used. News coverage from October 1979 showed that the 2-year-old dam was never once inspected, even though cracks had been reported. And there was no monitoring system in place to warn of spill threats. These shortcuts put an entire community in jeopardy.

At 5:30 a.m. on July 16, 1979—34 years to the date after the atomic bomb was tested in the New Mexico desert—the dam walls broke. Millions of gallons of contaminated water containing a range of toxic heavy metals flowed out into the Rio Puerco. Navajo lands far downriver flooded. Journalist Judy Pasternak described

Thousands of pounds of tailings are produced in uranium processing. It takes 1 ton of ore to extract 6 pounds (3 kilograms) of uranium, and the Church Rock mill processed about 4,000 tons per day. More than 3 million tons of tailings were stored in Church Rock's dam at the time of the accident.

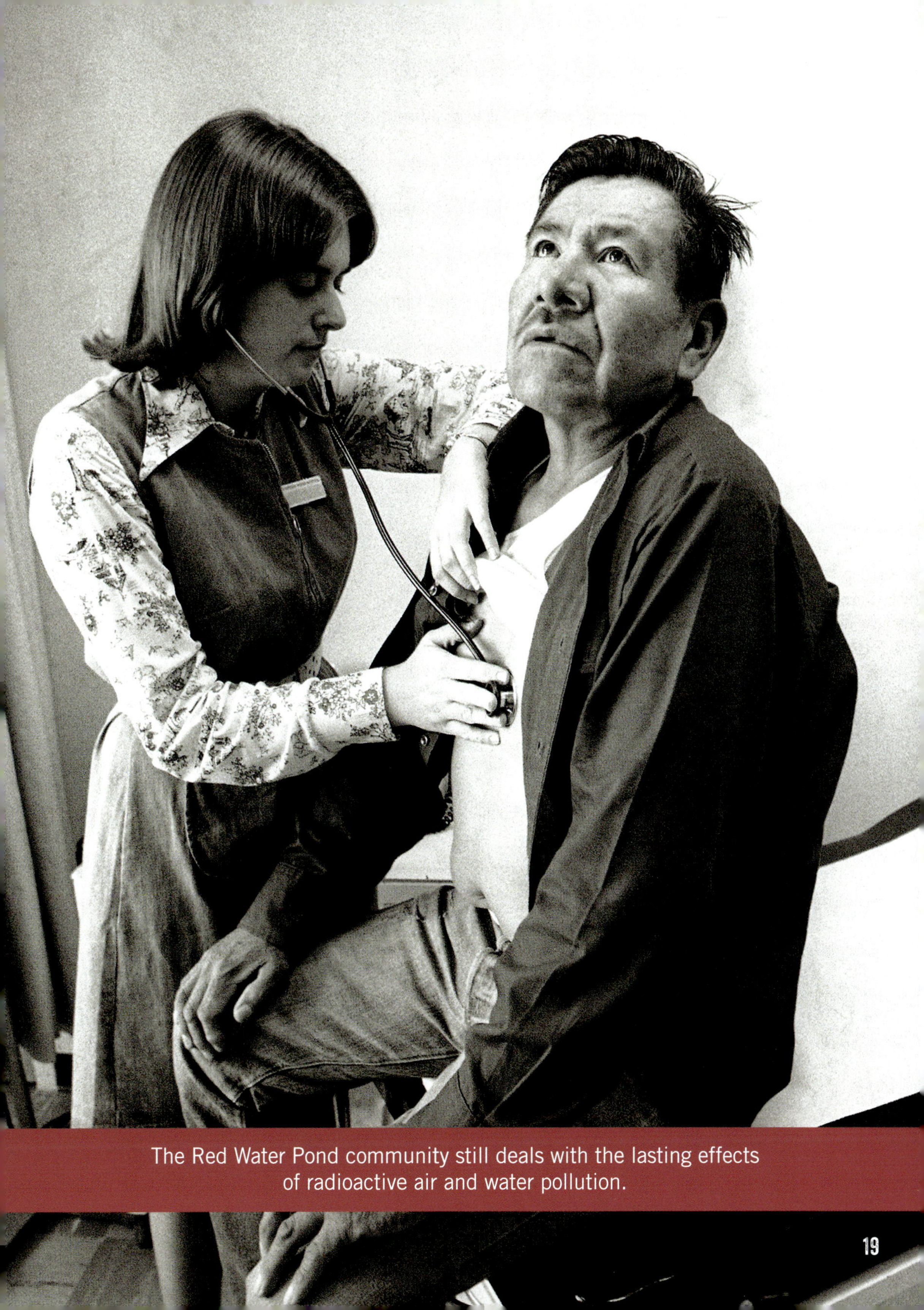

The Red Water Pond community still deals with the lasting effects of radioactive air and water pollution.

Unknowingly, both people and animals drank contaminated water less than a mile from the uranium mine.

the effects in her book, *Yellow Dirt*. Pasternak wrote, "Sheep keeled over and died, and crops curdled along the banks. The surge of radiation was detected as far away as Sanders, Arizona, 50 miles (80.5 km) downstream." Because the problem went beyond New Mexico and into Arizona, it became a multistate issue. At this time, the federal government should have intervened.

The worst nuclear catastrophe in history happened in 1986, when the Chernobyl Nuclear Power Plant in Ukraine melted down. Massive amounts of radioactive materials were released across the European continent. Because of widespread contamination, the site and the nearby city of Pripyat remain abandoned.

The Rio Puerco's waters no longer gave life, but destroyed it. Its **pH** measured around 1—safe pH levels for drinking water are generally considered to be between 6.5 and 8.5. The river itself was yellow, and the water was compared to battery acid. Despite this, a local news report published 2 days after the spill in the *Albuquerque Journal* quoted the head of New Mexico's radiation protection office offering assurances to citizens that "there is no immediate danger to people along the river or to wells up to Gallup." But humans and animals that waded into the river in the days and weeks afterward reported severe burns.

CHAPTER 3

After

It's now been more than 40 years since the uranium spill at Church Rock. In 2019, former mine surveyor and Navajo Nation resident Larry King talked to journalists at *Vice News*. King voiced frustrations common across the community. He said, "We have never been a priority. Forty years after the spill and nothing has been done." For decades, residents have worried about the things that other people take for granted, such as the right to clean air, soil, and water. And they worry about the health and well-being of their children and grandchildren. While they watched billions of dollars spent on cleanup efforts at Three Mile Island in Pennsylvania, their own suffering went largely unreported.

A former uranium mine worker shares his lingering health problems and his thoughts on the environmental injustices surrounding native communities with U.S. Representative Deb Haaland in 2019.

"Three Mile Island had more coverage and people were compensated right away. I always say we don't get the same attention because we live in impoverished native community. We live in a sacrifice zone." – Larry King, August 12, 2019 (Vice.com)

How did authorities go about cleaning up Church Rock's massive spill? After one month, the mining company had cleaned up about 50 of the 1,100 tons. But the way in which this was done—

Indigenous community leaders and environmental activists continue to push back on federal government interests in expanding uranium mining near the Grand Canyon.

with buckets, pails, and shovels, and without protective gear or heavy machinery—wasn't environmentally or socially responsible. It put those doing the work at great risk.

Further, no real urgency was attached to completing the work. This allowed the spilled toxins to seep deeper into the ground and the water table. Eighteen months after the accident, testing showed that radioactive materials and other heavy metals could be found at soil depths of 30 feet (9 m). In 1983, the mill and mine site was designated a **Superfund** site by the Environmental Protection Agency. Forty years later, cleanup work is still ongoing.

For the Navajo who have ancestral ties to the area, the spill created hard choices. Staying could endanger their health. But leaving could dissolve generations of cultural and community ties. From the beginning of the uranium boom, the population suffered high rates of certain cancers, autoimmune disease, and kidney disease. Infants were born with uranium in their urine. On top of that, residents experienced economic hardship. Ranchers couldn't sell livestock or meat that grazed on the land. Even people's homes became a threat: houses constructed from earthen material were found to have extremely elevated levels of radon and other hazardous gases in the air and dust.

Congress passed the Radiation Exposure Compensation Act in 1990. This piece of legislation provided money to former miners and those living downwind from nuclear test sites who developed health problems from the exposure.

Community members hand-make masks to help prevent the spread of the coronavirus on the Navajo Reservation.

With shifts in presidential administrations, cleanup efforts have varied in urgency. But Navajo activism has not. In 2005, the Navajo Nation banned uranium mining. Each year since 2009, July 16 has been marked as a day of remembrance. Marches, vigils, films, and discussions commemorate the spill's anniversary. In April 2016, activists appeared before the Inter-American Commission on Human Rights alongside other Americans subjected to **environmental racism**. They talked about their decades-long denial of the basic human right of clean water. And the community continues to hold monthly meetings to talk about solutions and the future. These activists are creating change—when and where they can—for a safer, cleaner, more just world.

In 2020, one-third of Navajo Nation residents had no running water at home. This had devastating consequences during the COVID-19 pandemic. Preventive steps like hand washing couldn't be easily taken. As a result, the Navajo faced one of the worst coronavirus outbreaks in the country.

Research & Act

In 1998, Lorissa Jackson, a high school teacher and member of the Navajo Nation, had an idea. Jackson realized that many of her students never learned about the history of uranium mining in their community. To help these young people see the effects of the mining industry on their families and the land, they decided to make a documentary film. The teens researched everything from the development of the atomic bomb to the lasting health impacts of uranium exposure. They interviewed tribal elders. They shared their film, *Hear Our Voices,* with fellow students, parents, and moviegoers in places as far away as Portland, Oregon.

Talking with people about their history offers a window into their experiences. Research a social justice struggle that impacted your community in the past. What effect does it have today? Learn about the issue. Make a list of questions that you have. Interview family members and neighbors, and, if you can, record your conversations. This will allow you to share their stories and perspectives with others, now and in the future.

Timeline

- **December 1938:** Nuclear fission is discovered, and government scientists race to develop an atomic bomb.
- **Early 1940s:** The U.S. government begins prospecting for uranium in the Southwest to use in nuclear weapons.
- **July 16, 1945:** The atomic bomb is tested in the New Mexico desert.
- **1968:** United Nuclear Corp. opens the Northeast Church Rock Mine on Navajo Nation land.
- **June 1977:** A uranium processing mill begins operating at the Northeast Church Rock site.
- **March 28, 1979:** An accident at the Three Mile Island Nuclear Power Plant in Pennsylvania sparks panic about nuclear safety.
- **July 16, 1979:** A massive dam break at the Church Rock Mine releases tons of radioactive waste into the Rio Puerco.
- **2005:** The Navajo Nation bans uranium mining.
- **July 16, 2009:** The 30th anniversary of the spill is commemorated with a march and vigil.
- **April 2016:** Navajo activists testify about water rights before the Inter-American Commission on Human Rights.

Further Research

Bruchac, Joseph. *Chester Nez and the Unbreakable Code: A Navajo Code Talker's Story.* Chicago, IL: Albert Whitman & Company, 2018.

Gilbert, Samuel. "Church Rock, America's Forgotten Nuclear Disaster, Is Still Poisoning Navajo Lands 40 Years Later." *VICE*, Aug. 12, 2019, www.vice.com/en_us/article/ne8w4x/church-rock-americas-forgotten-nuclear-disaster-is-still-poisoning-navajo-lands-40-years-later.

Kallen, Stuart A. *Navajo Code Talkers.* Minneapolis, MN : Lerner Publications, 2018.

Morales, Laurel. "For the Navajo Nation, Uranium Mining's Deadly Legacy Lingers." *NPR*, Apr. 10, 2016, www.npr.org/sections/health-shots/2016/04/10/473547227/for-the-navajo-nation-uranium-minings-deadly-legacy-lingers.

Native Planet. Performance by Simon Baker, season 1, episode 5, 2014.

Rangel, Valerie. "Church Rock Tailings Spill: 16 July 1979." *New Mexico History*, July 22, 2015, newmexicohistory.org/2015/07/22/church-rock-tailings-spill-16-july-1979/.

Glossary

assimilate (uh-SIH-muh-late) to make a smaller group similar to a dominant group's culture

chain reaction (CHAYN ree-AK-shuhn) a series of events that are triggered by one initial event

chronic (KRAH-nik) lasting for a long time

classified (KLAS-uh-fide) kept secret

Cold War (KOHLD WOR) the period of tension between the United States and former Soviet Union that lasted from about 1945 until 1990

dictator (DIK-tay-tur) a ruler with total power who typically governs by force

dosimeters (doh-SIM-ih-turz) devices that measure how much radiation is absorbed by the body

element (EL-uh-muhnt) a basic unit of matter

encrypt (en-KRIPT) to make into a code

environmental racism (en-vye-ruhn-MEN-tuhl RAY-siz-uhm) a form of discrimination often facing minority or low-income communities forced to live near environmentally hazardous sites

fallout (FAWL-out) radioactive particles that enter the atmosphere after a nuclear accident or explosion

Geiger counters (GYE-gur KOUNT-urz) tools used to detect radioactivity

nuclear fission (NOO-klee-ur FISH-uhn) the process of splitting an atom to create energy, first achieved in 1938

nuclear meltdown (NOO-klee-ur MELT-doun) an accident in a nuclear reactor in which the fuel overheats and melts the surrounding core and equipment

pH (PEE AYCH) the measure of the acidity or alkalinity of a substance

prospectors (PRAHS-pekt-urz) people searching for mineral deposits

radioactive (ray-dee-oh-AK-tiv) harmful atomic particles that are released when the nuclei of atoms break down

radium (RAY-dee-uhm) rare radioactive metal first discovered by Marie and Pierre Curie

reactors (ree-AK-turz) large devices in which nuclear energy is produced by splitting atoms under controlled conditions

Superfund (SOO-pur-fuhnd) a U.S. federal program set up in 1980 to pay for the cleanup of toxic waste

tailings (TAYL-ingz) the waste produced in the ore refining process

uranium (yu-RAY-nee-uhm) dense radioactive metal that is used as fuel in nuclear reactors

INDEX